PELORUS JACK

THE NEW ZEALAND DOLPHIN

WRITTEN BY
Darcy Pattison

ILLUSTRATED BY
Eva Dooley

Inspiring a Government to Protect an Individual Animal

Pelorus Jack, the New Zealand Dolphin: Inspiring a Government to Protect an Individual Animal

Mims House Books
1309 Broadway, Little Rock, AR 72202
MimsHouseBooks.com

Publisher's Cataloging-in-Publication data

Names: Pattison, Darcy, author. | Dooley, Eva, illustrator.
Title: Pelorus Jack , the New Zealand dolphin : inspiring a government to protect an individual animal / written by Darcy Pattison; illustrated by Eva Dooley.
Description: Includes bibliographical references. | Little Rock, AR: Mims House, 2024. |
Summary: The New Zealand government passed a law in 1904 to protect an individual dolphin, one of the first pieces of legislations worldwide to protect wildlife.
Identifiers: LCCN 2023921300 | ISBN: 9781629442419 (hardcover) | 9781629442426 (paperback) | 9781629442433 (ebook) | 9781629442440 (audio)
Subjects: LCSH Pelorus Jack (Dolphin)--Juvenile literature. | Wildlife conservation--New Zealand--Juvenile literature. | Dolphins--Juvenile literature. | Animals--Juvenile literature. | New Zealand--Juvenile literature. | BISAC JUVENILE NONFICTION / Animals / Marine Life | JUVENILE NONFICTION / Science & Nature / Environmental Conservation & Protection | JUVENILE NONFICTION / Science & Nature / Zoology
Classification: LCC QL737.C432 .P38 2024 | DDC 599.5/3--dc23

Chug, chug, chug. One day in 1888, a steamship sped through the water of Pelorus Sound, bound for the dangerous French Pass, and, beyond it, the city of Nelson, New Zealand. The steamer cut through the water, foam frothing at its bow. Suddenly, sailors saw a flash of white, something streaking toward them.

A mighty sea creature bounded over the waves, closer and closer. Suddenly, he vaulted high and struck the water with an enormous splash. About 12 feet long (3.6 meters), he frolicked, riding the bow waves, diving from side to side. He made the great steamer his playmate and fascinated the sailors.

Four steamers passed daily through the French Pass, traveling from Nelson to Wellington or back again. Soon the mysterious creature danced in the waves of every steamer crossing his six-mile stretch of Pelorus Sound. Delighted, the sailors named him Pelorus Jack.

Pelorus Jack met every steamer, day or night, rain or sunshine, year in and year out. As time passed, his fame grew.

Guidebooks talked about him...

visitors made time to travel to the French Pass area to catch sight of the fascinating creature...

and locals sold postcards with his photograph.

POST CARD

PELORUS JACK
FRENCH PASS, NEW ZEALAND

In 1895, American writer Mark Twain leaned on a steamer's railings, studying the strange creature.

In 1901, the young British royals, the Duchess and Duke of York (later crowned King George V), watched Pelorus Jack jumping through the waves.

In a 1902 issue of *New Zealand Illustrated Magazine*,
T. Lyndsay Buick wrote:

*"Exactly at the half-hour after midnight,
there came shooting out of the shoreward darkness
a comet-like stream of phosphorescent light...
'Pelorus Jack' darted right in front of the steamer...
as to color*

*he was simply
one blaze
of golden light...*

From the tip of his nose to the end of his tail
a mass of living gold...
racing and chasing,
"
leaping
and plunging."

One story says that a passenger aboard the *SS Penguin* took a shot at him, and afterward Pelorus Jack never swam with that steamer again.

Some reports said that European collectors were offering a bounty for Pelorus Jack's body.

It was clear that some ruthless people wanted to kill this amazing animal. Those who loved Pelorus Jack decided they needed the government to protect him. Early in 1904, the government agreed to help.

But what was Pelorus Jack? A fish? A whale? A dolphin?

For years, scientists had debated Pelorus Jack's species. At first, observers agreed he was a shark, but then most agreed he was a white beluga whale (*Delphinapterus leucas*). Based on eyewitness accounts, some scientists decided he was a Cuvier's beaked whale (*Ziphius cavirostris*), also known as the goose-beaked whale.

Perhaps, they thought, cameras could solve the question with solid evidence. But Pelorus Jack swam so fast that photographs only showed a blur. They even filmed him, but he was still a blur.

HECTOR'S DOLPHIN

DUSKY DOLPHIN

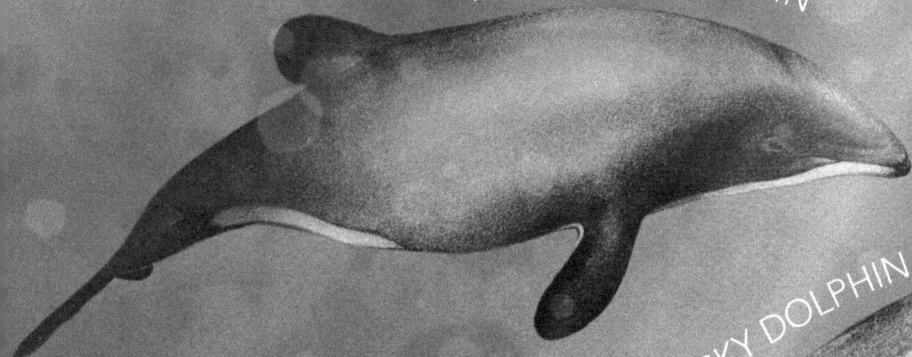

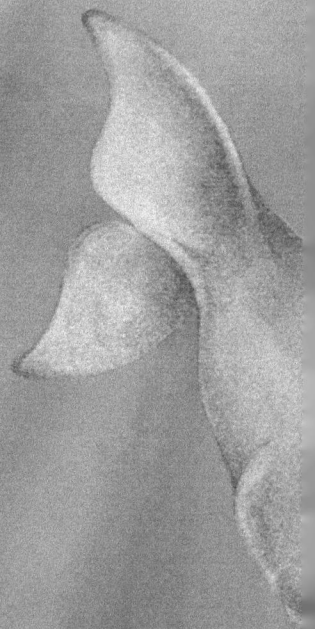

CUVIER'S BEAKED WHALE

BELUGA WHALE

In 1904, government official Reverend Daniel C. Bates reviewed all the eyewitness accounts and the photographs. The animal had a dorsal fin, which meant it wasn't a beluga whale. The blunt nose meant it wasn't a goose-beaked whale. Finally, Bates classified Pelorus Jack as a Risso's dolphin (*Grampus griseus*), a species not usually found in the New Zealand area.

On September 26, 1904, a New Zealand law finally protected the Risso's dolphin. It was the first time any national government had protected an individual animal.

One question remained: Why did Pelorus Jack swim with the steamers? Some reports said that the dolphin was piloting the steamers through treacherous waters. However, Pelorus Jack never went through the dangerous French Pass. He always stayed in the relatively calm waters of Pelorus Sound.

Maybe he just wanted to frolic in the bow waves.

Or maybe he enjoyed racing the steamers.

In the end, who knows the mind of a wild creature?

On March 20, 1911, sad news came. A large white marine animal had washed up on shore near the French Pass. Mr. Webber, postmaster of the French Pass, said it looked a lot like Pelorus Jack.

They contacted the captain of the SS *Pateena* to find out if he had seen Pelorus Jack that day.

On board the SS *Pateena*, the sailors hung over the railing, looking for their friend. For 23 years, the faithful dolphin had entertained them. Would he come again today? Or...

They scanned the waves for a white flash.

They worried.

They waited.

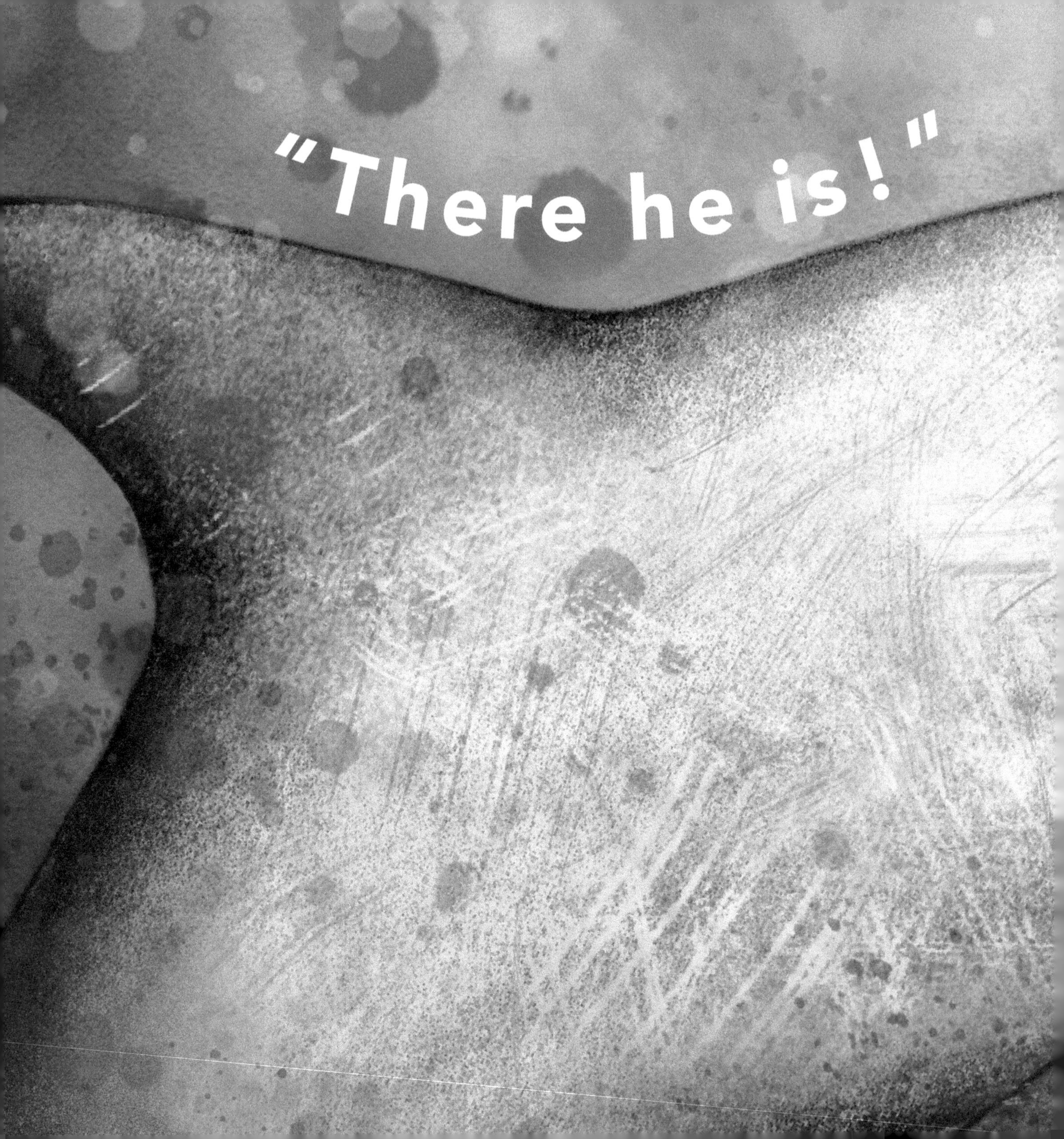

"There he is!"

Pelorus Jack was last seen in April 1912. Some sailors thought he had just moved on to different waters. Others suspected that Norwegian whalers shot Pelorus Jack to collect the bounty from European collectors. But most people think Pelorus Jack died of old age.

For 24 years, Pelorus Jack showed up to greet the steamers, day or night, rain or sunshine, year in and year out, a golden streak, racing and chasing, leaping and plunging.

He was a wild creature who inspired a national government to protect an individual animal.

Remembering Pelorus Jack today, people ask each other,

"How can we protect wild animals?"

Tall sharklike dorsal fin.

Single blow hole

Round head

Color ranges from dark gray to white. Scarring is common.

Teeth only in lower jaw.

8 to 13 feet (2.5 to 4 meters) long, weight up to 1100 pounds (500 kilos).

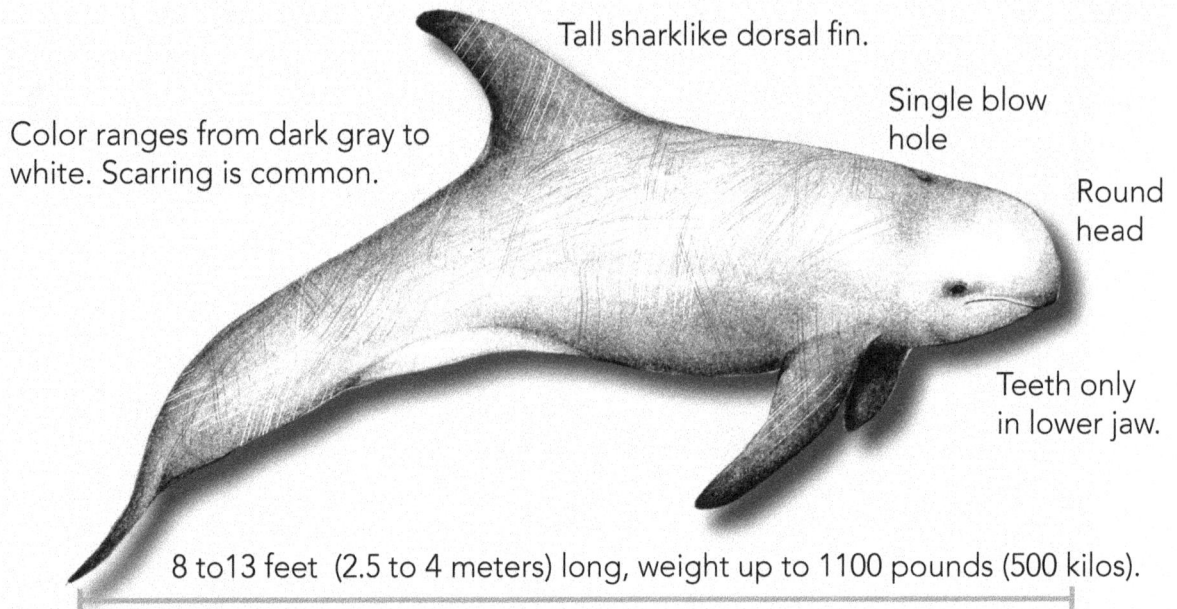

RISSO'S DOLPHIN (Grampus griseus)
Other names: Grampus; grey grampus; white-headed grampus; gee gee's

The Risso's dolphins live in tropical to temperate oceans around the world, although they are rare in New Zealand waters. Like all dolphins, they are marine mammals. Usually, they are seen in pods of 10 to 30 individuals. They have a large dorsal fin about halfway down their backs. Dark at birth, they become lighter as they age, and most Risso's dolphins have many scars on their skin. A crease runs vertically on their heads. They only have 4 to 14 peglike teeth in their lower jaw and no teeth in the upper jaw. To communicate, Risso's dolphins use a wide range of sounds, including clicks, barks, chirps, and whistles.

Risso's dolphins are very active on the water's surface, leaping, slapping their tails, and raising their heads out of the water, a move known as spy-hopping. They can dive 1000 feet (300 meters) deep and hold their breath for 30 minutes. Though they eat fish, they prefer squid. They feed at night, when their prey is active. Risso's dolphins are estimated to live about 35 years.

Though we call this famous New Zealand dolphin Pelorus Jack, we don't know if the dolphin was male or female. Male and female Risso's dolphins look very similar, and no one ever found Pelorus Jack's body to examine it.

PELORUS JACK. This remarkable white fish, a species of dolphin, accompanied for several miles all steamers passing through the French Pass, on the Nelson-Picton run, New Zealand. "Jack" was estimated to be from 12 to 16 feet in length, and was protected by Government regulation.

MĀORI LEGENDS

Pelorus Jack was known as "Kaikai-a-waro" by native New Zealand Māori. Their stories say that the dolphin was a sea god who saved the lives of early Polynesian people who arrived in New Zealand by canoe.

SOURCES

Papers Past. National Library of New Zealand. Multiple newspapers from 1904 to 1915 with mentions of Pelorus Jack. paperspast.natlib.govt.nz.

Buick, T. Lyndsay. "Pelorus Jack." *New Zealand Illustrated Magazine*, volume VII, issue 3 (December 1, 1902), page 210. https://paperspast.natlib.govt.nz/periodicals/NZI19021201.2.13

"The French Pass and Pelorus Jack," *Marlborough Express*, volume XLI, issue 56 (March 7,1907), page 7. Franck T. Bullen's comments about Pelorus Jack. https://paperspast.natlib.govt.nz/newspapers/MEX19070307.2.40

Lawson, Will. "Pelorus Jack: A Complete History of the Wonderful Pilot Fish of New Zealand." *Pacific Marine Review: The National Magazine of Shipping*, volume 21 (September 1924), page 459. https://books.google.com/books?id=RT8fAQAAMAAJ

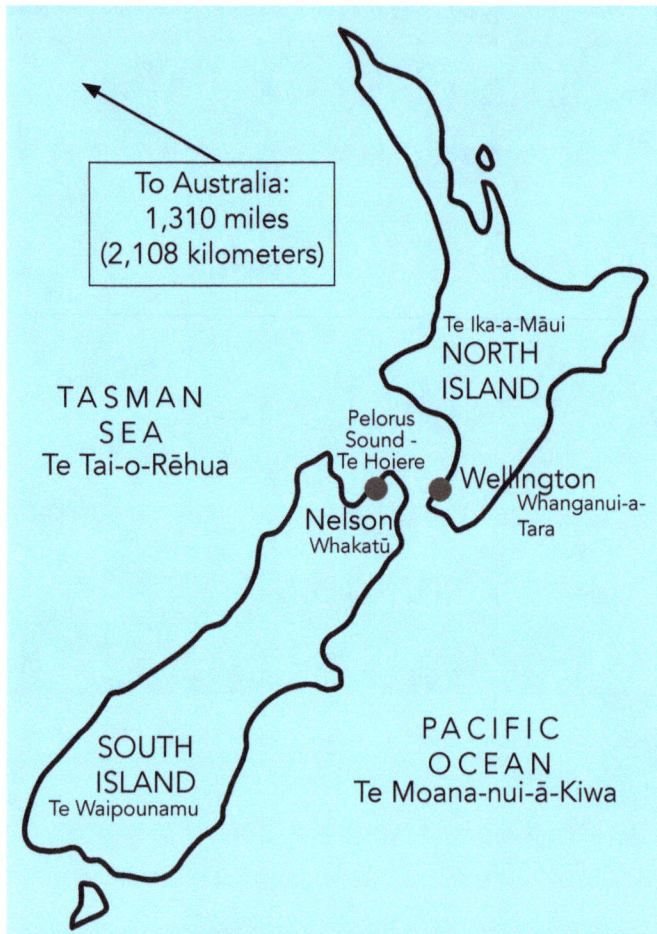

To Australia:
1,310 miles
(2,108 kilometers)

Te Ika-a-Māui
NORTH
ISLAND

TASMAN
SEA
Te Tai-o-Rēhua

Pelorus
Sound -
Te Hoiere

Wellington
Whanganui-a-
Tara

Nelson
Whakatū

SOUTH
ISLAND
Te Waipounamu

PACIFIC
OCEAN
Te Moana-nui-ā-Kiwa

NEW ZEALAND

New Zealand (Aotearoa, in the native Māori language) is an island country in the southwestern Pacific Ocean, with two main islands and over 700 smaller islands. Its closest neighbor is Australia, to the west. The capital is Wellington (Whanganui-a-Tara), and Auckland (Tāmaki Makaurau) is the largest city.

Originally settled by Polynesians, New Zealand developed a distinctive Māori culture. In 1769, British captain James Cook explored and mapped the New Zealand coastlines. In 1841, New Zealand became a colony of the British Empire. By 1947, it had become an independent country.

Communication between the North Island (Te Ika-a-Māui) and the South Island (Te Waipounamu) has always been important. Pelorus Jack was named after the Pelorus Sound (Te Hoiere), a body of water at the northern tip of the South Island. Travelers going from Wellington, on the North Island, to Nelson (Whakatū), on the South Island, traveled through Pelorus Sound.

When steam-powered ships replaced sailing ships in the mid-1800s, the steamship became the most common mode of transportation between the islands. It's said that each ship's motor has a distinctive sound, and that Pelorus Jack could tell them apart.

Left: Moetapu Bay, on the Inner Pelorus Sound, looking north towards the Outer Pelorus Sound. Photo by Karora. Image in the public domain. https://en.wikipedia.org/wiki/File:Moetapu_Bay-Inner_Pelorus_Sound.jpg

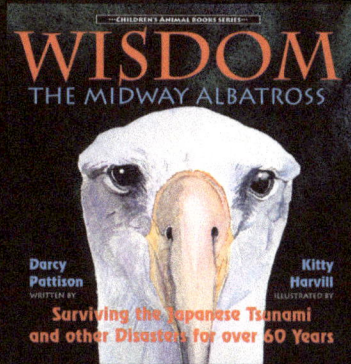
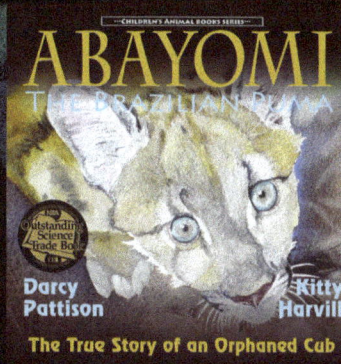
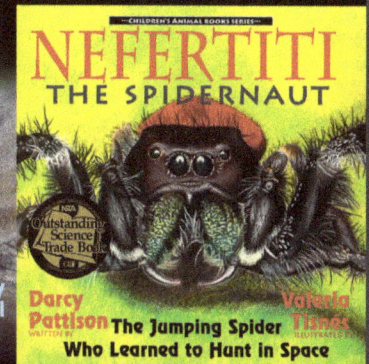
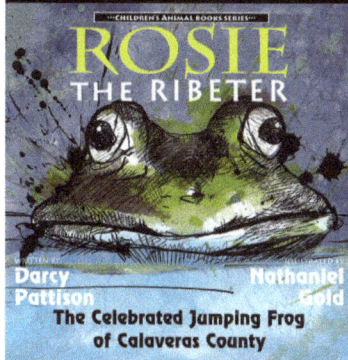
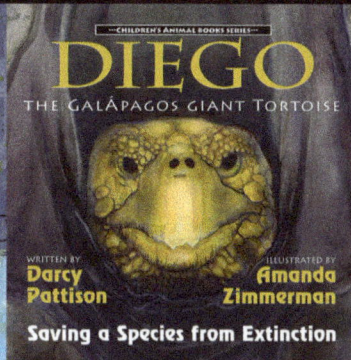
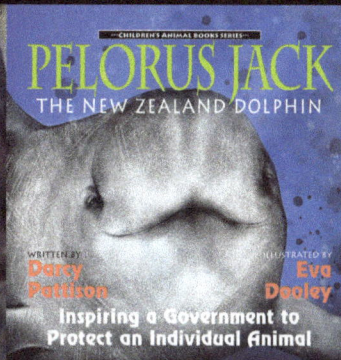
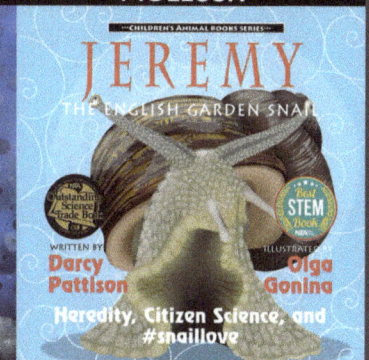